SPACE MYSTERIES

HOW WILL PEOPLE TRAVEL TO MARS?

Gareth Stevens
PUBLISHING

BY EMILY MAHONEY

Please visit our website, www.garethstevens.com. For a free color catalog of all our high-quality books, call toll free 1-800-542-2595 or fax 1-877-542-2596.

Cataloging-in-Publication Data

Names: Mahoney, Emily.
Title: How will people travel to Mars? / Emily Mahoney.
Description: New York : Gareth Stevens Publishing, 2019. | Series: Space mysteries | Includes index.
Identifiers: LCCN ISBN 9781538219539 (pbk.) | ISBN 9781538219515 (library bound) | ISBN 9781538219546 (6 pack)
Subjects: LCSH: Space flight to Mars--Juvenile literature. | Mars (Planet)--Exploration--Juvenile literature.
Classification: LCC TL799.M3 M34 2019 | DDC 523.43--dc23

First Edition

Published in 2019 by
Gareth Stevens Publishing
111 East 14th Street, Suite 349
New York, NY 10003

Copyright © 2019 Gareth Stevens Publishing

Designer: Katelyn E. Reynolds
Editor: Joan Stoltman

Photo credits: Cover, p. 1 MARK GARLICK/SCIENCE PHOTO LIBRARY/Getty Images; cover, pp. 1, 3–32 (background texture) David M. Schrader/Shutterstock.com; pp. 3–32 (fun fact graphic) © iStockphoto.com/spxChrome; p. 5 Steve Allen/DigitalVision/Getty Images; p. 7 NASA/JPL-Caltech/University of Arizona/Texas A&M University; pp. 8, 17, 22, 23 NASA; p. 9 NASA/JPL-Caltech/Dan Goods; p. 11 NASA/JPL; p. 13 NASA, ESA, and The Hubble Heritage Team (STScI/AURA); p. 15 NASA/JPL-Caltech; p. 18 NASA/Pat Rawlings, SAIC; p. 19 NASA/JPL-Caltech/MSSS; p. 21 NASA's Goddard Space Flight Center, Phobos and Deimos images courtesy of NASA/JPL-Caltech/University of Arizona; p. 25 NASA/SebiPhys/Wikipedia.org; p. 26 Brendan Hoffman/Bloomberg via Getty Images; p. 27 Marc Ward/Stocktrek Images/Getty Images; p. 29 Steven Hobbs/Stocktrek Images/Getty Images.

All rights reserved. No part of this book may be reproduced in any form without permission in writing from the publisher, except by a reviewer.

Printed in the United States of America

CPSIA compliance information: Batch #CS18GS: For further information contact Gareth Stevens, New York, New York at 1-800-542-2595.

CONTENTS

Our Neighbor Planet 4
Ice Cold! 6
First Sightings 8
Exploring the Surface 10
Failed Missions 12
Gathering Information 14
People on Mars? 16
Radiation Dangers 18
Asteroid Assistance 20
A Deep Space Gateway 22
Deep Space Transport 24
One Big Step 26
A Future on Mars 28
Glossary 30
For More Information 31
Index 32

Words in the glossary appear in bold type the first time they are used in the text.

OUR NEIGHBOR PLANET

For thousands of years, people have looked up at the night sky and wondered what life might look like on other **planets**—if life even exists there at all!

Perhaps the most interesting planet to us is Mars because it's Earth's neighbor. While we have sent many spacecraft to Mars, no person has ever been there. Recently, humans have made it their **mission** to send a person to Mars to learn more about this strange, red planet.

OUT OF THIS WORLD!

Mars is about 140 million miles (225.3 million km) away from Earth, on average.

OUR SOLAR SYSTEM

Neptune
Uranus
Saturn
Jupiter
Mars
Earth
Venus
Mercury
Sun

Mars is the fourth planet from the sun.

5

ICE COLD!

Part of the reason people have never been to Mars is because its **environment** isn't great for humans. It's very cold and dry. The only water on the planet is found in the clouds and icy soil.

If people were to go to Mars, they would need to wear special safety clothing to keep them warm and help them breathe. They'd also need a way to collect water. There are many things that make traveling to Mars hard!

OUT OF THIS WORLD!
Mars is the name of the Roman god of war. The Romans gave the planet Mars its name because of its blood-red color!

The surface of Mars looks very different from Earth!

7

First Sightings

In 1965, the spacecraft *Mariner 4* flew past Mars and sent 22 pictures of Mars back to Earth. These were the first closeup pictures ever taken of another planet!

People loved looking at these pictures, and the idea for a journey to Mars started to take shape. Since Mars can be reached within a person's lifetime, the chance to travel to another planet seemed possible. However, many things would need to be considered before a trip to Mars could happen.

Mariner 4

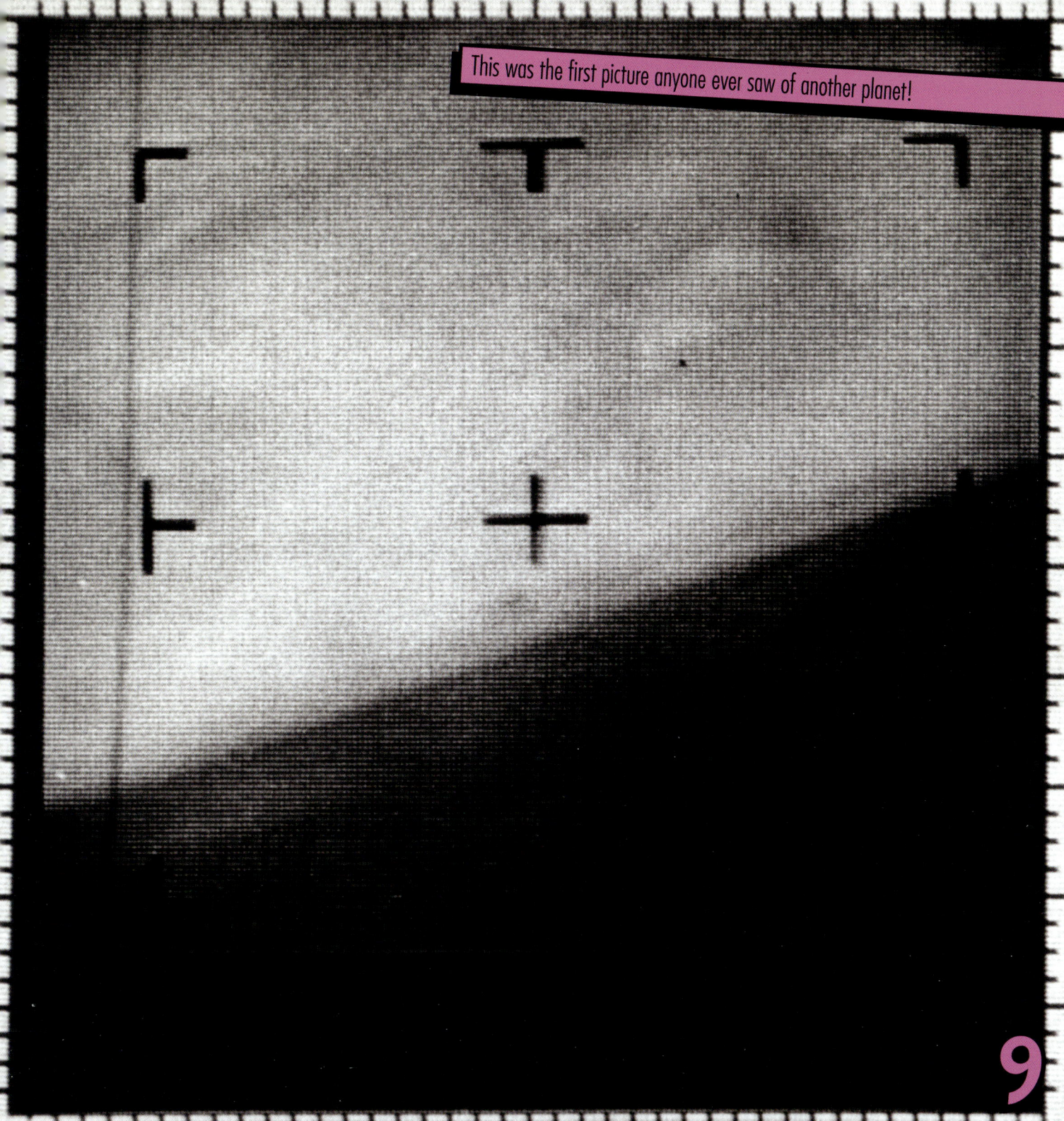

This was the first picture anyone ever saw of another planet!

9

EXPLORING THE SURFACE

In 1976, spacecraft called *Viking 1* and *Viking 2* reached the surface of Mars. However, it wasn't until 1997 that a **rover** called *Sojourner* actually drove around on Mars and explored the planet's surface. *Sojourner* operated for about 3 months and collected a lot of important data.

It was clear that machines could land on Mars. With each new Mars mission, more data pointed to the possibility of finding water on Mars. The idea of people visiting the planet grew.

OUT OF THIS WORLD!

July 4, 1997—the day *Pathfinder* landed and began sending pictures of Mars back to Earth—has been called "the day the internet stood still" because so many people were looking for pictures online.

The rover *Sojourner* traveled on a larger spacecraft called *Pathfinder*. Pathfinder was **launched** on December 4, 1996, and landed on Mars on July 4, 1997.

Sojourner

Pathfinder

11

Failed Missions

Sending spacecraft into space can be dangerous. Two more spacecraft were sent to Mars in 1998 and 1999, but they were lost upon arrival. Scientists think one of the spacecraft may have burned up when it got close to the planet.

Failed missions are terrible because of the huge amounts of time and money spent for each. A failed mission with people traveling aboard a spacecraft to Mars would be even worse. Scientists are planning carefully for possible future manned missions to the planet.

Mars's thin atmosphere—the gases around a planet—is made mainly of a gas called carbon dioxide, which is deadly to humans in large amounts.

13

GATHERING INFORMATION

Gathering data is an important step in sending people to Mars. Scientists want to understand Mars as much as possible so they can keep **astronauts** safe.

Another rover is set to travel to Mars in 2020. This rover will collect rock and soil samples. Tests run by the rover will hopefully give scientists clues about how astronauts could use the planet's **natural resources**. The rover will also attempt to turn carbon dioxide into oxygen, a gas humans need to breathe.

OUT OF THIS WORLD!

There are sometimes huge, long-lasting dust storms on Mars. The 2020 rover mission will collect data about the planet's dust.

This is what the 2020 Mars rover may look like!

15

PEOPLE ON MARS?

NASA believes it may be possible to send people to Mars by the 2030s. They're already preparing for this historic trip by sending rovers and other spacecraft to explore the planet.

Scientists and astronauts need to think about more than just how to get people safely to Mars. They also have to figure out how astronauts will **orbit** the planet, survive in Mars's atmosphere, and, of course, return home to their families. As you can imagine, this is quite the task!

OUT OF THIS WORLD!

Spacecraft have taken between 128 and 333 days to travel from Earth to Mars depending on how far away the planets were from each other on the launch date.

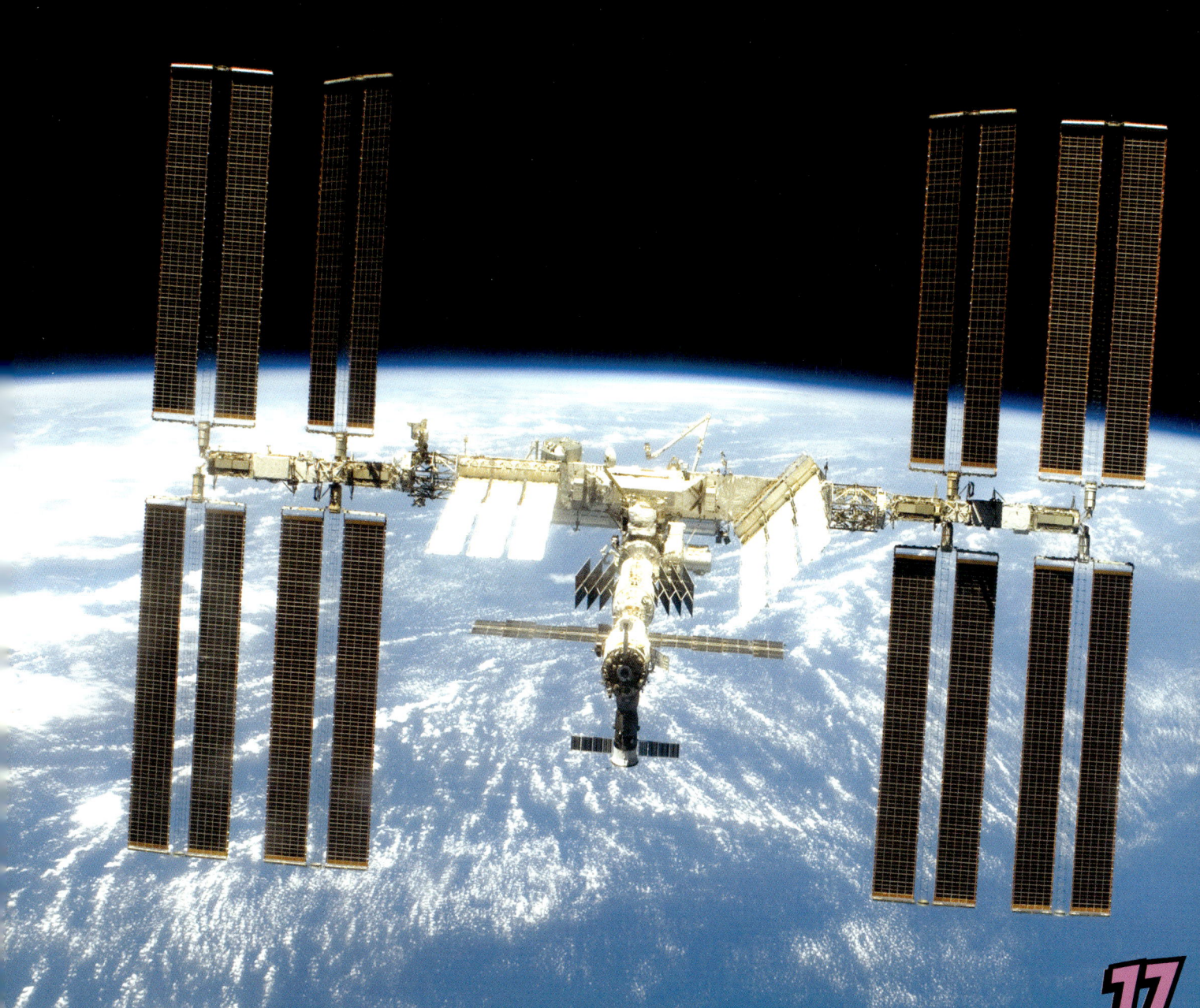
The International Space Station (ISS) will one day be the first step on a trip to Mars!

RADIATION DANGERS

The journey to Mars is about more than just building a spacecraft. Scientists must also consider the weather on Mars and in space! For example, solar, or sun, storms can produce a lot of radiation. Radiation is a type of **energy** that can't be seen or felt. It's very dangerous to people and animals!

NASA is currently designing spacecraft **shields** to block radiation. They're also creating drugs that may be more effective than shields at **protecting** astronauts from certain types of radiation.

Scientists are very hopeful that **future** exploration of Mars will include humans.

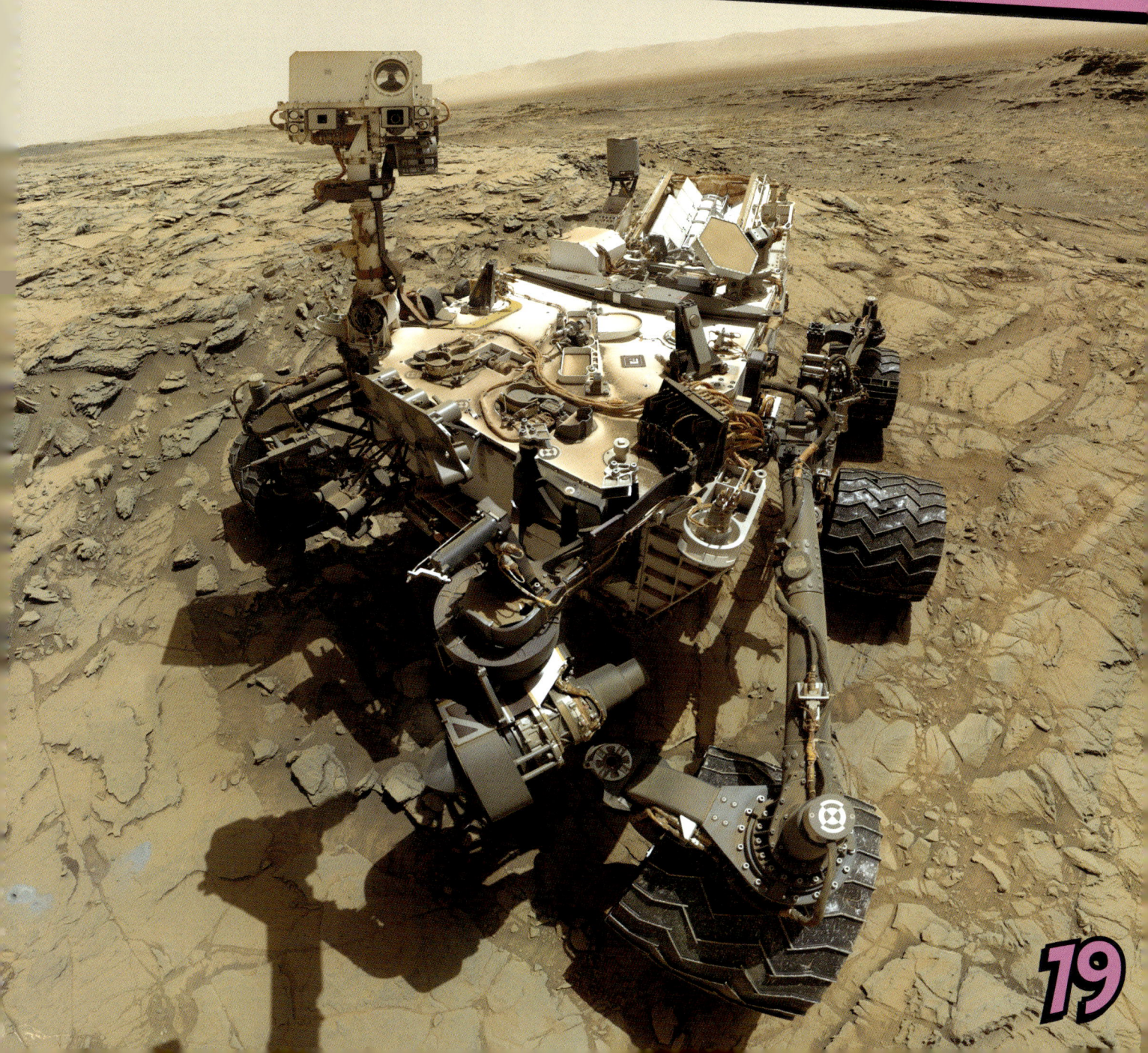

Curiosity, a car-sized rover, measured the amount of radiation in space as it traveled to Mars in late 2011. As of November 2017, it was still in operation, measuring radiation on the planet's surface.

ASTEROID ASSISTANCE

One of NASA's next steps is a mission in deep space. During this mission, scientists hope to capture an **asteroid** and move it into the moon's orbit. Asteroids can help scientists learn about sending people to Mars.

NASA hopes astronauts on the *Orion* spacecraft will be able to collect samples from the captured asteroid in the 2020s. This mission will help NASA test new systems that may be used to send supplies to Mars before human missions begin.

OUT OF THIS WORLD!

Mars's moons may actually be asteroids that were pulled into Mars's orbit because of a force called gravity. Gravity pulls objects toward each other.

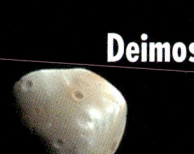
Deimos

Phobos

Mars's two moons—Phobos and Deimos—were discovered in 1877.

A DEEP SPACE GATEWAY

NASA is planning to build a new space station near Earth's moon. This station will act as a deep space gateway to the moon's surface. It will also provide **support** for spacecraft and astronauts traveling deeper into space, such as on future missions to Mars.

Before building a deep space gateway, NASA is sending the *Orion* spacecraft into space aboard the Space Launch System (SLS)—the world's most powerful rocket. This is the first of several missions to test systems needed for deep space travel.

SLS

Orion, the SLS, and the deep space gateway will give astronauts important experience for deep space travel. They will also allow astronauts and spacecraft to depend less on Earth when on deep space missions.

Mars

asteroids

Earth's moon

Orion

Earth

DEEP SPACE TRANSPORT

NASA is trying to figure out how to safely transport, or move, people into deep space. They're working on a deep space transport spacecraft they hope will allow people to travel to Mars in the near future.

The deep space transport will be able to be used again and again, returning to the deep space gateway to be reloaded and repaired, or fixed. The deep space transport spacecraft will be tested near Earth's moon to be sure it will hold up on longer missions into deep space.

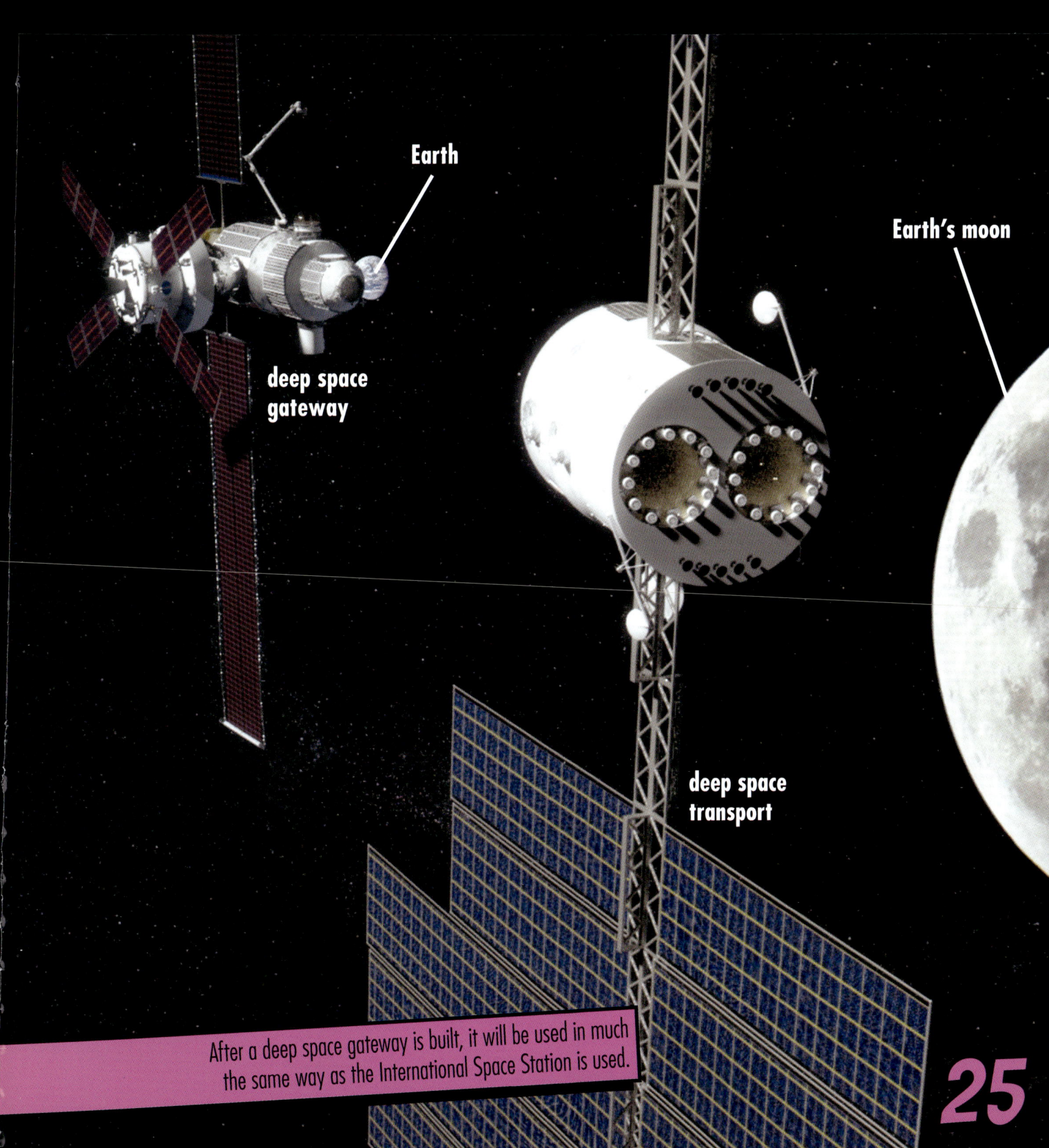

After a deep space gateway is built, it will be used in much the same way as the International Space Station is used.

25

ONE BIG STEP

NASA is looking to send astronauts into orbit around Mars by 2033. Based on the success of this mission, they'll work on their plan to land people on Mars. However, other companies, such as SpaceX, have said they plan for a human to walk on Mars in the early 2020s.

Whoever makes it to Mars first will have the special opportunity to do something no human has done before—walk on a planet other than Earth!

Elon Musk, CEO of SpaceX

People could be walking on Mars sooner than we think!

A FUTURE ON MARS

As scientists learn more about Mars and its atmosphere, traveling there in the near future is looking more and more possible. Scientists, astronauts, and even ordinary people are becoming very excited by the idea of traveling to Mars and maybe even living there some day!

Who knows? One day you may be getting up in the morning, putting on your space suit, and traveling to work in a Mars rover. Until then, keep your eyes on the night sky.

OUT OF THIS WORLD!

Other than Earth, Mars is the planet with the best conditions for supporting life.

Glossary

asteroid: a large rock that orbits the sun

astronaut: someone who works or lives in space

energy: power used to do work

environment: the conditions that surround a living thing

future: the period of time that comes after the present

launch: to send out with great force

mission: a task or job a group must perform

natural resource: something in nature that can be used by people

orbit: to travel in a circle or oval around something, or the path used to make that trip

planet: a large object in space that moves around a star

protect: to keep safe

rover: a small, moving machine used for exploring the surface of a moon or planet

shield: something that protects someone or something

support: to hold up and help

FOR MORE INFORMATION

BOOKS

Aldrin, Buzz, and Marianne J. Dyson. *Welcome to Mars: Making a Home on the Red Planet*. Washington, DC: National Geographic, 2015.

Bailey, Diane. *The Future of Space Exploration*. Mankato, MN: Creative Education, 2013.

Hartman, Eve, and Wendy Meshbesher. *Mission to Mars*. Chicago, IL: Raintree, 2011.

WEBSITES

Everything Mars
kids.nationalgeographic.com/explore/space/everything-mars/
Visit this website to play games and find fun facts about Mars and outer space.

Mars for Kids
mars.nasa.gov/participate/funzone/
Learn more about Mars through the fun activities on this website.

NASA Photojournal
photojournal.jpl.nasa.gov/
Find amazing images of Mars and other planets here.

Publisher's note to educators and parents: Our editors have carefully reviewed these websites to ensure that they are suitable for students. Many websites change frequently, however, and we cannot guarantee that a site's future contents will continue to meet our high standards of quality and educational value. Be advised that students should be closely supervised whenever they access the internet.

INDEX

asteroid 20, 23

astronauts 14, 16, 18, 20, 22, 23, 26, 28

atmosphere 13, 16, 28

carbon dioxide 13, 14

Curiosity 19

deep space 20, 22, 23, 24

deep space gateway 22, 23, 24, 25

deep space transport 24, 25

dust storms 14

Earth 4, 5, 7, 8, 10, 16, 23, 25, 26, 28

Earth's moon 20, 22, 23, 24, 25

gravity 20

International Space Station (ISS) 17, 25

Mariner 4 8

Mars's moon 20, 21

Musk, Elon 26

Orion 20, 22, 23

oxygen 14

Pathfinder 10, 11

radiation 18, 19

rover 10, 11, 14, 15, 16, 19, 28

Sojourner 10, 11

solar storms 18

solar system 5

Space Launch System (SLS) 22, 23

SpaceX 26

Viking 1 10

Viking 2 10

3 1333 04780 7696